A New Approach to Evolution

Acknowledgements

I am most grateful to Professor Alan Holland, Professor Emeritus of Philosophy, University of Lancaster and to Professor Sir David Weatherall FRS of The Weatherall Institute, the University of Oxford for criticism and advice.

To Ruth, Wendy and Ann

Copyright © 2011 Richard Sheriff Jones

The right of Richard Sheriff Jones to be identified as the Author of this Work has been asserted by them in accordance with the Copyright, Designs and Patents Act 1988.

ISBN: 978-0-9559017-8-2

www.p3publications.com

First Published in Great Britain
in October 2011
by P3 Publications,
13 Beaver Road,
Carlisle,
Cumbria,
CA2 7PS

Printed and bound
in Great Britain
by Amadeus Press
Cleckheaton
BD19 4TQ

Contents

Chapter		Page
1	Philosophy and Evolution	4
2	Old and New Views of Evolution	7
3	Understanding the Minome	11
4	Evolution	13
5	The Emergence of Consciousness	16
6	Human Understanding	18
7	The Minome and Evolution	20
8	Twenty First Century Philosophy	23
9	Conclusion	29
	Gene Consciousness	30
	Theory of Mind Implications	32
	The Evolution of Philosophical and Psychological Ideas	35
	The Gene Mover	39
10	Glossary	40
11	Index	43

1. Philosophy and Evolution
Humanity's search for truth

We think we are in the driving seat, yet the road we travel is not in ourselves but in the Gene Mover.

R Sheriff Jones

      Philosophy is defined here as a subject in which the knowledge content is from understanding based on 1) sensory experiences of the universe and 2) the subjective responses to these experience, called here 'the Sensibilities for Feelings, Aesthetics and Values. *Both the sensory experiences and the responses to these are unknowable.* The reasons for this statement and the implications of it will be discussed. But from this foundation it is possible to establish a philosophical description of human understanding and its limitations that are free from common errors of logical origin. What follows shows that some essential philosophical facts about understanding have been ignored in the past. The main thrust of this work is to establish how human understanding came into being during the course of evolution and its significance for life in the 21st century.

The New Approach

Understanding is a word that is commonly used but what does it mean? It appears to be central to thinking and to all knowledge. On analysis it quickly becomes evident that it is multifactorial, so that it is impossible to supply a simple definition.

Over the last three centuries philosophers developed an analytical system that sought to explain the universe in the light of accumulating scientific knowledge. Much of it was valuable because it explored the limits of human understanding. But by the end of the 19th century it ceased to have clearly defined objectives. Yet most of the post-modernist philosophers that followed have continued to use the analytical method to follow ever more closely what scientists are doing and to interpret their findings.

There is at present no accurate and agreed description that defines the terminology to be used for knowledge acquisition and therefore there is no generally accepted definition of what is meant by 'understanding' and 'meaning'. This has led to errors and these are compounded, as will be shown, when examining evolution in animals and humans. Richard Sheriff Jones pointed out that the correct terminology must be based on data obtained by introspection and this is described in detail in *Beliefs and Human Values*, (BHVS, Dialogues 6-13) [1] and, in summary, is set out below.

What has not been appreciated is that the introspective method of analysis clearly and accurately defines the details and extent of human understanding: It explains the relations between scientific knowledge, the normative ethical values of society and faith-based beliefs. Introspection has been ignored because it could not be validated by reason. Modernists such as Ryle[2], did not accept introspective evidence, but that was an error left over from the Age of Reason. What is now required is not the application of more reason, but an accurate description of all aspects of understanding. Everyone is aware of their own private components of consciousness. They are amenable to accurate description and are not accessed by reasoned philosophical interpretation. The crucial importance of this method is shown here where it will become apparent how understanding evolved in evolutionary and genetic terms.

---

[1] *Beliefs and Human Values*, Richard Sheriff Jones 2009, Trafford ISBN 978-1-4251-7313-5 (sc) and 978-1-4251-7314-2 (e).
[2] *The Concept of Mind*, Gilbert Ryle, 1949 Published by Hutchinson and Co, Ltd.

Evolution is currently discussed in terms of the biological gene and changes in the phenotype that relate to Darwinian survival of the fittest. But the terminology in which these are discussed is pre-rather than post-modernist. Philosophers accept current scientific knowledge but many still use reason in the manner of the pre-modernists to construct a description of current evolutionary knowledge. The contention here is that this terminology is out of date and often leads to incorrect conclusions.

The objective is to describe the origin of understanding in terms of evolutionary genetics. I begin with a description of what is meant by the terms 'understanding' and 'meaning'. Understanding is incomplete in early childhood when the brain is developing and decreases when there is dementia and becomes disorganised in mental disease. But this article is not concerned with these events.

Sensory organs, including vision and therefore consciousness, developed at an early stage in evolution, more than 130 million years ago, but may be much older. This I call the beginning of the Brain Age and with it came the Ability to understand. All understanding is in response to sensory experiences of objects moving in time and space. Although time and three dimensional space are essentially unknowable, it is from these that scientists since Greek times have used reason to construct hypotheses in order to understand or explain the universe. Currently the universe is thought to have begun with the Big Bang and has since been expanding. Although such hypotheses are the products of our awareness of time and space, they do not explain the nature of time and space, nor do they explain such questions as the extent of time and space in the universe as described by scientists today. These products of reason supply limited knowledge and therefore understanding from sensory experience. But clearly this knowledge has no significance except when it has relevance to human requirements.

2. Old and New Views of Evolution

Since there is no accurate and agreed description at present that defines the terminology to be used for knowledge acquisition, there is no generally accepted view of what is meant by 'understanding' and 'meaning'. This has led to errors and these are compounded, as will be shown, when examining evolution in animals and humans. Let us begin therefore by examining the elements on which understanding is based:

1. <u>Perception:</u> the reception of sensory experiences by the brain and their passage into consciousness

2. <u>Reason,</u> which is of two kinds, as is explained in the conclusion to this paper. αReason denotes reason applied to sensory experiences of the inanimate universe and to animate beings within it, which includes all the sciences. In the sciences, reason makes understanding possible and operates in conjunction with logic and mathematics to increase accuracy. βReason is reason directed subjectively in *response* to sensory experiences and is applied to the Sensibilities of Feelings, Aesthetics and Values.

3. <u>The Sensibilities</u> are integral parts of mind in Homo sapiens but are present in lesser and variable degrees in nonhuman primates. In fact, Sensibility differences are the only way at present in which it is possible to measure differences of understanding between humans and other Brain Age organisms. Discussed below.

4. <u>Concept formation.</u> Conceptual Ideas enter consciousness continuously during waking hours and it is from these that the construction of human understanding becomes possible. This is a large and complex subject. Most of these concepts do not survive and are discarded. Some become beliefs and of these a small number come to form a part of established knowledge. Rarely, this knowedge is from concepts that have never previously existed (as in science or music) and are called proto-concepts.

5. <u>The Memory</u> is required to maintain all of these functions, which in part explains the size of the human brain. Without memory, understanding is not possible (as in early infancy and dementia).

6. <u>Abilities</u> are the hidden drivers of all mental and physical activity.

These components are responsible for *all* the elements of human understanding[3] and together I call them the *Minome* because they possess special properties described in BVHS. When the Abilities listed are analysed, it becomes clear that our only conscious awareness of them is solely *when they are in use:* we only know them indirectly through the medium of examples. Therefore the Minome is a list of those unknown or unknowable properties of which everyone is aware, yet they are only knowable in terms of examples in use. After much tribulation, philosophers have accepted now that all knowledge has to be expressed in terms of examples and not in terms of an imaginary objective dimension representing a reality that was assumed must exist.

Some people find it difficult to accept that these components of understanding are in some sense hidden so that the concept of the Minome they find difficult. The philosopher and mathematician Bertrand Russell explained it in this way [4]. 'Pure mathematics was discovered by Boole, 1854. We start in pure mathematics from certain rules of inference, by which we can infer that if one proposition is true, then so is some other proposition. These rules of inference constitute the major part of formal logic. We then take any hypothesis that seems amusing and deduce its consequences. *If* our hypothesis is about *anything*, and not about one or more particular things then our deductions constitute mathematics. Thus mathematics may be defined as the subject in which we never know what we are talking about, nor whether what we are saying is true.' Thus Russell illustrates very clearly that the tools we have available do not enable us to understand anything; that reason alone contributes nothing towards understanding. But if I ask 'what does 2+2 tell us? ' the answer is immediately clear. And this mysterious hidden quality characterises all the elements of understanding, which together I call the Minome.

We know that Brain-Mind relations have evolved from a simple yet remarkably sophisticated level in the ant, for example, to the human level. During this period, brain size in humans increased dramatically relative to body size and it increased in absolute terms during the later stages of the Brain Age in H. sapiens. The latter phase was probably in response to the demands of the later stages of Minome formation (Reason and the Sensibilities).

---

[3] The word 'Understanding' includes ideas of cause and effect, purpose, teleology (explanation by reference to an end or purpose). But always understanding is completely dependent on the accuracy of inanimate and animate sensory experiences.

[4] *Mysticism and Logic* Bertrand Russell, Published by Pelican 1918 p74/5.

It is not surprising that the brain size had to increase greatly during evolution to sustain this complexity of understanding. It must have increased in response to gene-based stimuli but the mechanism is at present unknown. Also unknown is how the elements of understanding listed above evolved over millions of years. The Feeling Sensibilities must have appeared first, while the Aesthetic Sensibilities appeared later and the Value Sensibilities relatively recently. These are all questions to which biological scientists such as Lewis are seeking answers. In 1992 E. B. Lewis reported the discovery by himself and others of a cluster of control genes, BX-C that act to enhance or suppress the target genes that bring about the differentiation of tissues and organs of the third thoracic and abdominal segments in the phenotype[5]. These have been found in insects as well as humans. BX-C may once have been part of a larger regulator cluster, ANT-C on the same chromosome. He adds that there is evidence also that proteins bind these regulator genes and this may explain why there is evidence that the cluster, HOM-C, has remained linked for over 500 million years, which is before the separation of vertebrates from invertebrates. A H Knoll [6] provides a useful summary of the earliest forms of life. He agrees with the suggestion of C. R. Woese that major episodes in eukaryote evolution correspond with major episodes in the earth's geological history[7].

But what is the role of the philosopher? For millennia prior to Greek philosophy, humans had depended for their understanding of life and death on faith-based beliefs. But the advent of Greek science and Platonic philosophy showed that an alternative conceptual belief system was possible. Science has since become an entrenched part of knowledge about the universe, but for over a century the role of the philosopher has become questionable. D.C. Dennett [8] writes on page 2 of his book, 'Freedom Evolves',

'We are each made of mindless robots and nothing else, no non-physical, non-robotic ingredients at all'.

And on page 15,

'My fundamental perspective is *naturalism*, the idea that philosophical investigations are not superior to, or prior to investigations in the natural sciences, but [are] in partnership with those truth seeking enterprises, and that

---

[5]Clusters of Master Control Genes Regulate the Development of Higher Organisms, *Edward B Lewis JAMA March 18th, 1992, Vol 267No 11.*

[6]*The Early Evolution of Eukaryotes: A Geological Perspective,* Andrew H Knoll 1992 Science Vol 256, pages 622-627

[7]C R Woese, *Microbiol Rev (1987) 51 221.*

[8]*Freedom Evolves,* D.C. Dennett, Published by Penguin Books 2003.

the proper job for philosophers here is to clarify and unify the often warring perspectives into a single vision of the universe. That means welcoming the bounty of well-won scientific discoveries and theories as raw material for philosophical theorising, so that informed and constructive criticism of both science and philosophy is possible'.

Is this statement correct? While accepting the scientific method, Dennett envisages a role for the philosopher, which is in the tradition of analytical philosophy prior to the 20th century. Richard Rorty, trained as an analytical philosopher, struggled but never managed to escape entirely from the limitations of that philosophy[9]. When a scientist undertakes research, he follows an untrodden path to an uncertain end, where hopefully he is able to form an hypothesis that when verified will become knowledge. But the philosopher is unable to assist in this because he does not possess, as the scientist does, a great deal of prior knowledge in his field, often accumulated over many years. Philosophical discussion of problems that may arise take place *after* this process, not during it.

Dennett goes on to present a determinist/materialist theory of consciousness, but adds 'I encounter pockets of uneasiness, a prevailing wind of disapproval or anxiety quite distinct from mere scepticism'. When details of the elements of understanding listed above are carefully examined, there is no possibility that a determinist/materialist explanation of understanding could accord with the unknowns that characterize sensory experience. The origins of all *objective* sensory experiences and the origins of all *subjective* responses to these experiences by the various Sensibilities *are both unknowable*. These facts have to be accepted to make Dennett's position understandable. It is surprising how often the unknowable components of existence are ignored, which serves to compound the extent of the errors that result.

Dennett entitles his book *Freedom Evolves* and I agree that there is no end point to the evolution of Conceptual Ideas that appear and may come to form beliefs. But the motivation to form new beliefs usually comes only when existing beliefs are no longer thought to be true, or of use. The most important reason for this is that sensory experiences of both the inanimate and animate worlds keep changing. True freedom would only be possible should beliefs about sensory experiences cease to form, but then human existence would cease. That is not the way brain-mind relations work. Beliefs can however be minimized, as shown in BHVS, Part 6, which reduces the dangers they may cause.

---

[9]*Philosophy and the Mirror of Nature*, Richard Rorty, 1979 Princeton Press.

3 Understanding and The Minome

Each of the 6 elements of understanding listed above is capable of bringing about its own deadly sin if one is not careful! We are clearly aware of each of these functions in consciousness, yet the *awareness* of them is always in terms of examples. We are only aware of mathematics on inspection of a specific equation; or of good and evil when an example of something judged to be good or evil is experienced. But we are not aware of what brings about, or causes these judgements. Therefore, *all* those elements of understanding that are unknowable I have described collectively as the 'Minome', the significance of which will become clearer later. The elements of the Minome are made up of all the objective and subjective phenomena that are unknowable. As a consequence of these facts, I find it necessary to define philosophy so that it *includes* those aspects of the objective and subjective unknowable components of understanding.

Hence it is evident that *the boundaries of philosophy extend beyond those of science* and the implications of this fact are profound. However, they do not appear to be widely appreciated by philosophers. Arguably the most important conclusion in all philosophy is that *it is precisely from things unknowable that humans discover the ideas, which they regard as most valuable.* But scientists simply ignore things unknowable, for these are not required to achieve the purely *tentative* conclusions of science.

Clearly, the role of the philosopher must incorporate these facts, although there will always be some who accept beliefs that restrict this role, as Dennett does, and others who extend them in a wide variety of ways by means of faith-based beliefs. M Mameli attempts to defend Dennett's position[10] when he claims that *'philosophical storytelling, if it is properly informed by up to date scientific knowledge, can still be useful'.* He describes four possibilities but they are not convincing because three are from the science of psychology and are therefore about science, not philosophy. However, his fourth point is relevant and important.

---

[10]On Dennett and the Natural Sciences of Free Will Matteo Mameli, a review of Daniel C Dennett, Freedom Evolves, Allen Lane, Penguin Press, London 2003. See also Comment on Dennett, Consciousness Explained BHVS, Discussion 14

Einstein states the problem here very clearly[11] and philosophers are in a powerful position to convey scientific results to the public. But I find the terminology in Mameli's paper unsatisfactory: 'the buck stops here', the 'source of responsibility intuition'. What precisely do these terms mean and what do the terms 'will' and 'intuition' really mean? He accepts *physicalism*, (the principle that all explanations of whatever is claimed to be knowable, must reduce to physical explanations of the way the brain works). But this is incompatible with the fact that there are aspects of the universe that are undeniably unknowable.

Philosophers in the 21st century, however, are in a potentially powerful position, as will be found later for, on the one hand they remain true to the facts of science as these evolve, and on the other they are always responsive to changing human values within societies. Both are demanding because science is so often misunderstood, terrorism is rampant and because religion and the family doctor, which used to be the bedrocks of support in society, have dropped away.

The terminology in biological and philosophical literature is not systematised and is often borrowed from different branches of science, philosophy and everyday life. The contents of the Minome (listed on p 4) describe only what is observable. References to intellect, mind, person, personality, and soul are unsatisfactory unless qualified.

---

[11] "Anyone who has ever tried to present a rather abstract scientific subject in a popular manner knows the great difficulties of such an attempt. Either he succeeds in being intelligible by concealing the core of the problem and by offering the reader only superficial aspects or vague allusions, thus deceiving the reader by arousing in him the deceptive illusion of comprehension; or else he gives an expert account of the problem, but in such a fashion that the untrained reader is unable to follow the exposition and becomes discouraged from reading any further. It is not sufficient that each result be taken up, elaborated, and applied by a few specialists in the field. Restricting the body of knowledge to a small group deadens the philosophical spirit of a people and leads to spiritual poverty." Quoted in the Frontispiece to BHVS.

4 Evolution

Having described the elements of understanding, the obvious next question to ask is where did understanding come from and how did it evolve during the course of evolution? An army of scientists is currently investigating the mechanism of evolution in a variety of ways. The main thrust of progress in the 20th century was to place on a scientific footing the gene and the Darwinian concept of survival of the fittest.

Before the Brain Age, evolution may be said to have been 'wild' in the sense that, the gene simply explored every niche available to an organism in a given physical environment, which was rapidly being filled with the organisms of other species competing with each other for survival. The gene created an infinite variety of heritable species, but many did not have the fitness to survive. It seems probable that this phase of genetic wild card evolution had largely run its course during the Brain Age by the time that larger brained animals began to appear. Did brains come into existence because the limits to simpler pre-Brain Age forms of life creation were being reached? From the evidence available, it would appear that at least two factors might have come into play.

1. First, the possibilities for pre-Brain Age kinds of life creation were becoming exhausted.
2. Secondly, a much more effective and powerful method of evolution had appeared in the form of animals with brains. Bigger brains were required to operate the new Minomic functions that had been developing over millions of years.

Virtually nothing is known about these gene-based events. One must assume that brains did not grow spontaneously, but in response to sensory stimuli from organs that were appearing on the surface of certain animals. How did the evolutionary gene know that this would lead to what I call the Brain Age that has proved to be an extremely powerful method for fostering life on earth? In our current ignorance of relations between gene and phenotype, P E Griffiths and E M Neumann-Held drew a useful distinction between two gene concepts, the 'Molecular gene': the entire molecular process involved, whatever that turns out to be, and the 'Evolutionary gene': a theoretical entity that includes phenotypic traits. These need not correspond to specific stretches of DNA [12].

---

[12] *The Many Faces of the Gene* 1999, Paul E Griffiths and Newman-Held, Bioscience Vol 49 No 8:656-662.

There is no simple relationship in the present state of knowledge between the molecular gene and the phenotypic traits that it manifests. The relations between these have been difficult to determine but these authors have reviewed the possibilities. Population genetic theories are central to neo-Darwinism and were formulated before examination of genetic material became available. In view of the current scientific uncertainty, the authors suggest the provisional distinction between what for simplicity they call the 'gene' or *Molecular gene* and its phenotypic manifestations, the *Evolutionary gene*. These uncertainties must have a bearing on Darwinian fitness to survive and must correlate with Minomic functions, for the latter may explain some of the current discrepancies, which will be explained below. In view of these terminological difficulties, I use the word 'gene' in this article to refer to those elements of the genome, as yet not definable, that exist in each individual.

J T Bonner discusses possible reasons for the appearance and increase of multicellular organisms in early evolution[13]. C H Waddington noted that some phenotypic variation occurred from generation to generation but without modification of the genotype. He explained this by introducing the term 'canalisation'[14]. It had been thought that genotype determines phenotype, but by the seventies, this proved to be too simplistic. Patrick Bateson reviewed the literature at that point[15]: 'By the early 1970s Ethology itself was ripe for takeover. Its Grand Theory was in ruins and the much hoped for understanding of the links between behaviour and underlying mechanisms was still fragmentary'. Bateson and Mameli discuss the possibility that some phenotypic characteristics can be distinguished as 'innate' as distinct from others that are 'acquired', but concluded that there is no evidence to support such a distinction in population studies[16]. See also Mameli and Bateson who arrive at the same conclusion in respect of the Sciences.[17] Stuart A Newman is strongly of the view that extragenetic cell activity could have been responsible

---

[13] The Origins of Multicellularity, 1998 John Tylor Bonner: Integrative Biology. Vol 1 p27-36

[14] Canalisation of Development and the Inheritance of Acquired Characteristics 1942 C H Waddington Nature: No 3811 pp563-565

[15] The promise of behavioural biology 2003, Patrick Bateson: Animal Behaviour, 65, p11-17

[16] The Innate and the Acquired: Useful Clusters or a Residual Distinction From Folk Biology2007, Patrick Bateson and Matteo Mameli, Dev Psychobiol 49:818-831

[17] Innateness and the Sciences 2006 Matteo Mameli and Patrick Bateson, Biology and Philosophy, 21: 155-188.

for many of the phenotypic results observed during evolution.[18] Stuart A Newman and Gerd B Müller discuss these Epigenetic Mechanisms[19].

As neuroscience has progressed, brain-mind relations and behaviour have become important. G Gottlieb made an extensive study of behavioural genetics and concluded that the standard use of population statistics is too inaccurate. To enhance accuracy it must be replaced with studies of individuals over generations (he used mice mainly and followed changes of behaviour over up to 39 generations)[20]. Such studies must be regarded as very preliminary attempts to arrive at something remotely resembling elements of the Minome observed in the ant, which is described below. Gottlieb [21] presents a model (Gottlieb Fig 2.2) that includes a partial description of the sensory-motor axis that I described (BHVS and below, p 14) showing the relations between physical (sensory) experiences and behaviour (motor) experiences. The model treats genes as an integral part of a developmental system, which fits well with the concept of consciousness presented below. What is missing from his model is a description of the Sensibilities and how they relate to all the other functions of the Minome.

---

[18]*Developmental Mechanisms: putting genes in their place.2002 Stuart A Newman, J Bioscience, Vol 27 no 2 pp 97-104*

[19]*Epigenetic Mechanisms of Character Origination 2000, Stuart A Newman and Gerd B Muller: Journal of Experimental Zoology 288: 304-317*

[20]*On making Behavioural Genetics truly Developmental 2003, Gilbert Gottlieb: Human Development, 46 pp 337-355*

[21]*Chapter 2 from book by Byron C Jones and Pierre Mormede Neurobehavioural Genetics, Methods and Applications.*

5 The Emergence of Consciousness, the Trajectory and the degree of perfection of Gene products

In the early pre-Brain Age, sensory organs began to appear on the surface of some animals. These increased the effectiveness and therefore fitness of these organisms to survive. Among these was the eye, which caused a further decisive advance, which I have defined as the beginning of the Brain Age that commenced in the Precambrian era. This development brought consciousness into existence and with it mental Abilities of increasing complexity, summarized below. Human understanding is the product of these Abilities. During the *Brain Age*, the gene continued to explore possibilities under the sea and in the air, but the culmination of these developments was on land. Alongside these gene-based evolutionary changes with their phenotypic manifestations, mental Abilities within consciousness came into existence. The status of these Abilities has not been decided. It is claimed here that the gene has produced these Abilities within consciousness, but their manifestations are different from gene activity outside of consciousness. These differences and their implications will now be described.

'Survival of the fittest' is an incomplete description. When the gene produces a new species, it survives for a time but all species ultimately become extinct although the objective is always survival. *Each species sets out on a trajectory* that determines to a considerable degree its own fate. Subsequent genetic and epigenetic changes may produce new modifications and thence new species with their own trajectories and levels of fitness to survive. But the monkey and any modified offspring of monkeys are not on trajectories that will allow those species to evolve into humans, for example.

In the case of the ant, its structures and functions reached *a level of perfection* that proved to be remarkably durable. When this happens, there can be no other end point except the formation of sub-species, of which about 12,000 have formed over the 130 million years of the ant's existence. Having reached this point of 'perfection', the options left are survival unchanged until for example, extinction occurs on account of destruction by a virus, or disappearance of the fungus on which ants feed.

Each species, whether birds in flight, monkeys climbing trees, or fish hunting, appear to reach a state of perfection in respect of the skills that have evolved and presumably this position has always been the evolutionary endpoint because it represents maximum survival possibilities for the fittest species. And when a particular skill has been exploited to the utmost, the species must ultimately die out. Its very survival success appears to seal its fate. This principle appears to have operated prior to the Brain Age as well as during it. But during it and within the Minome, the as yet immature Feeling Sensibility guides the sensory-motor activities that drive evolution forwards, as described below.

The sensory-motor axis appears to operate within and outside the Minome on the basis that small changes in gene function occur over millennia and become manifest as changed phenotypic features, better adapted to survival. Within the Minome, such changes become manifest as changes in human physical and/or mental Abilities, which is presumably what has happened during the Brain Age over millions of years. Humans do not appear yet to have reached any definitive end-point of 'perfection'. Is that because the Minome has become so powerful that human objectives now dominate over extra-Minomic genetic effects? Yet humans are still at the mercy of hidden and unexpected viral mutations, for example, as well as by threats due to man-made atomic explosions, or by allowing the global population of humans to outstrip the resources of the planet to sustain them.

But apart from the gene in evolution, the *principle of perfection* applies also within the lifetime of the individual. All motivation is directed towards maximising objectives of one kind or another and these continuously evolve. Also, nations strive towards similar goals in democracies as well as when ruled by tyrants etc. It is significant that at the Minomic level of gene function, the crucial decisions that determine the fate of humans are all ethical, hence the importance of Reason plus the Sensibilities in the philosophy of this subject as discussed below. This subject must surely be one of the dimensions of human affairs that will become one of primary concern to philosophers in the 21st century?

One may ask: what other evidence is there for gene control of the Minome? The evidence is in the strengths and weakness of the Abilities that are available to each individual. Although these are in the Minome, their manifestations are wholly dependent on the animate (which includes other humans) and the inanimate environment (the physical universe). There are plenty of questions for philosophers and psychologists to answer. They include the following: will new Sensibilities slowly appear? Is the Value sensibility changing, as suggested by the increasing emphasis during the last few centuries on benevolence? Why does this fail to evolve in some individuals indicating that they lack a sense of values? Is reason becoming a more powerful Ability, or are we simply finding new technical ways of applying it?

6 Human Understanding

It appears probable that understanding evolved to reach its current level of maturity in humans during the last 130,000 years. But the emergence and the level of maturity reached by each of the Abilities described (primarily Reason/Logic/Mathematics plus Conceptual Ideas from sensory experience of the universe; and Reason, Logic and Mathematics plus the Sensibilities and Responsive Conceptual Ideas from the subjective domain of consciousness) probably increased slowly over millions of years during the Brain Age. The Ability levels (the drivers for Conceptual Ideas achieved for each function in every individual also seems likely to have varied, which would explain the marked differences in Ability levels that exist today within communities and between individuals in each family. But the possibility of obtaining evidence of these changes during the evolution of Brain Age animals is problematic.

R W Byrne has made an extensive study of the evolution of primate cognition[22]. He concludes that humans are descended from an African Great Ape clade and divergence occurred about 30 million years ago. The simian branch then developed larger brains. Byrne argues that in later stages of the Brain Age, absolute brain size is a better indicator of cognitive power than brain relative to body size. This is likely to have been when Reason plus the Sensibilities were nearing their current level of development. Evidence from Neanderthal man with a brain size equal to H.sapiens is meagre but would be valuable.

In H. sapiens today, the gene *determines well-defined limits* available to each individual for each function (to do mathematics, win races, or compose music, etc). But in addition, and perhaps even more important, there is a wide variation in the degree to which the various Abilities that actually exist in each individual are put to use to achieve objectives, (some individuals are lazy and do not make use of their Abilities and others among these may become 'down and outs'. A minority are very active and creative so that some may become leaders of professions, or of communities, and a few become dictators). Within communities differences do not correlate with the Ability to accumulate knowledge (the winner of the BBC brain of Britain competition may earn a living as a taxi driver). Many differences await investigation. Are differences due to the presence or absence of Abilities, or lack of opportunities to use them, or lack of the motor impulse to achieve? How are such differences to be measured?

---

[22] *Evolution of Primate Cognition* 2000 Richard W Byrne: Cognitive Science, volume 24 (3), pp 543-570

At the end of the article quoted above (footnote 9), Mameli correctly states that it is very likely that some non-genetic inheritance mechanisms have a high degree of evolutionary significance'. Again I stress that these apparently non-genetic mechanisms are within the Minome and therefore must be regarded as within gene control, in the sense explained further. The gene appears to have become manifest in ways that are different within the Minome from what they were in the pre-Brain Age, as explained below. In addition to the examples of Ability differences mentioned above, it must be assumed that such differences are likely to have applied throughout the Brain Age and may have affected the course of evolution in a variety of ways.

S J Gould and R C Lewontin wrote an article that in many ways goes to the heart of what it means to understand in scientific terms [23]. When the hypothesis was accepted that the biological gene and survival of the fittest were the explanation for heredity, it was assumed by many that all phenotypic traits observed must have been adapted to subserve this objective. But the authors point out that Darwin himself vigorously asserted that other as yet unknown factors must be assumed to contribute to the phenotypic traits observed. In other words, one hypothesis must not be allowed to exclude other possibilities. Hence, the hypothesis of the Minome and its properties presented here must be considered on its merits.

The formation of Conceptual Ideas describes the continuous stream of ideas that enter consciousness continuously and are discarded or accepted as tentative beliefs, which in turn may become items of knowledge. There are an infinite number of distinctive Conceptual Ideas but those that become beliefs and comprise knowledge are broadly of four kinds,

1) The tentative beliefs of science from sensory experience of the inanimate universe, which can be measured more accurately than the other categories.
2) The tentative beliefs of individuals about the animate world, which can be measured, but less accurately.
3) The normative beliefs of communities, which are pragmatic and keep changing. They can also be measured, but again less accurately.
4) The faith-based beliefs of religion, atheism etc, that are not amenable to 'proof' of any kind.

Reason and the Sensibilities are involved in *all* knowledge acquisition. There exist an indefinite number of Feeling, Aesthetic and Value Sensibilities. It is essential to appreciate that members of each of the three Sensibilities are involved

---

[23] *The Spandrels of San Marco and the Panglossian Paradigm: a critique of the adaptationist programme* 1979 Proceedings of the Royal Society. London B 205, 581-598

in every decision made by humans.[24] The motivating agent in this system I have called the *sensory-motor axis* (defined further below), because this has been the unique principle common to all forms of life from unicellular organisms up to and including the Brain Age and the Minome. The sensory component is from sensory experiences and the motor component from the unknowable cause that I call the Abilities. The specific Abilities of each individual seek objectives (in the form of ideas and actions that may follow). The Abilities are products of the gene in consciousness.

The *Minome* is the prime mover for all activities initiated in consciousness. To learn is to accumulate knowledge, which means to store it in the memory and use it for specific purposes. Memory also performs another function: as life advances towards its end a trail of memories remains, even when the memories of individual experiences have largely been forgotten. As Minomic functions deteriorate, the remaining memories increase greatly in value, for in the end they are all that is left.

Learning is a complex process and in the human is likely to utilise all of the functions of the Minome at any one time. To say that learning is an alternative method of inheritance implies that it has nothing to do with the gene. That would make the Minome an independent agent, which would be an error because Minomic functions are wholly dependent upon brain-mind functions. Hence what is in consciousness is subject to modified gene effects.

Much of the current confusion arises because scientists in different fields (such as psychology, biology and genetics) each develop their own different terminologies. In addition, philosophers try to amalgamate and add to these. But neither philosophers nor the laity should attempt to predict the ways in which an individual science will develop and the terminology it will require to do so, as this would risk teleological conclusions and/or metaphysical blind alleys. The end-point of a scientific theory can never be predicted.
Scientists investigating brain-mind relations will solve many of these apparent problems, so that they will simply disappear. Mameli discusses in *Mindreading, Mindshaping and Evolution* [25] some aspects of evolution and genetics. Recently it has been realised that learning, imprinting and similar habits observed in a wide variety of species may bring about hereditable results (produced, it is assumed by the gene plus other methods of transmission),

---

[24] How these complex relationships operate to form judgements and arrive at objectives is described in BHVS, R. Sheriff Jones.

[25] Mindreading, Mindshaping and Evolution, 2001 Matteo Mameli, Biology and Philosophy 16: 597-628

which have recently been summarized by Avital, E and Jablonka, E[26] and by Mameli.[27] Mameli confined his attention in this paper to vertical transmission from parents to offspring. He avoided horizontal transmission (transmission within a community, for example) because this creates the problem that it is thought to be difficult to know what is gene determined and what is determined by learning and similar habits. And one reason why this has caused confusion is because it entails knowledge of the Minome of which a description was not available. The consequence is that 'learning' has tended to be regarded as some kind of floating (non-genetic) asset available to humans when required. But learning requires all the elements of the Minome and therefore is an integral component of human understanding, *which means it is under gene control*. Mameli assumes that phenotypic changes due to learning etc would eventually be 'genetically assimilated'. The evidence for that is problematic, but if the Minomic hypothesis is correct, no assimilation is required for *the changes observed are already within gene control in the Minome*. As pointed out below, in humans no gene mutations appear to occur that produce supermen or superwomen. Observed mutations are all failures to achieve normality. This strongly suggests that the gene control of consciousness and the uses humans make of Minomic functions are very effective as explained below.

    Mameli considers inheritance by psychological and sociology methods and uses 'tools' (reasons based on arguments) to arrive at his conclusions. For example, he uses reason applied to the term 'psychology' to arrive at certain conclusions. But 'Psychology' is a term that applies only to humans because it draws on the full range of Reason plus the Sensibilities to attain meaning. It ceases to have meaning applied to animals in the Brain Age such as the butterfly and the ant, which possess only a fraction of the human range of Sensibilities. Similar arguments apply to sociology. These terms cannot be applied to animals with very different evolutionary trajectories and to human Minomic functions without making serious errors. Philosophers often invoke the use of 'tools' to explain some aspects of evolution, but the only tools available are from the Minome. There are in fact a wide variety of conceptual beliefs that are often called tools. But they perform different tasks in each of the four very different operations that are outlined above.

---

[26] *Animal Traditions Behavioural inheritance in Evolution*, Avital E and Jablonka E, 2001 Cambridge University Press.

[27] *Nongenetic Selection and Nongenetic Inheritance*, Matteo Mameli, British Journal for the Philosophy of Science 2004: 55 p35-71

7 The Minome and Evolution

It was noted above that huge differences exist between individuals in the total knowledge they store and there are dramatic differences in the degree to which individuals and societies utilise their Abilities. But much greater differences exist in these categories of knowledge between H. Sapiens and other animals. These differences are as yet mostly anecdotal and have not been established on a scientific basis. However, in the animal kingdom learning is an important evolutionary mechanism. Learning is a gene dependent function, contrary to what is sometimes claimed. It is a word that draws upon most or all Minomic functions in the human. Only rudiments of Reasoning plus some Feeling Sensibilities are the functions involved in the learning activities of ants for example. Elements of the Aesthetic Sensibility are evident in bird song and the sexual ritual of mating in humming birds must be regarded as rudimentary evidence of the Aesthetic Sensibility, which is much more highly developed in H sapiens. But this is an area where a great deal of work has yet to be done. Are there grounds for making any statements at all about the learning capacities of larger animals, for the evidence seems uncertain? Lack of progress appears to be because emphasis has been placed on comparing the Ability to reason in animals with human Abilities whilst the Sensibilities are ignored.

Among the smallest and simplest of brains must be those of bees and ants. Bees build hives. There are queens, workers and drones. There is considerable evidence of social relationships, coordination and cooperation, all of which must relate to survival. Social life has some special significance. Humans must also live in societies or die. The need for social life must have special significance. It suggests that Minomic functions could only have developed in a social setting and when one refers back to those functions it becomes obvious that this must be so.

The ant is one of the earliest Bain Age animals and has been in existence for millions of years. When an ant's nest is examined, are all the activities observed meaningless? Or is it possible to understand what they do, and possibly why, in terms of the understanding of humans today as described above? The facts from lay as well as scientific sources show that a great deal of human understanding can be applied to answering these questions.

'Go to the ant, thou sluggard, consider her ways, and be wise'. [28]
The Old Testament prophet had obviously made careful observations. He must have studied ant colonies and recognised activity that reminded him of ideas with

---

[28] Proverbs chapter 6, verse 6 The King James version of the Holy Bible.

which he was familiar in the society of his day.

Now consider the scientific evidence. Ants descended, as did bees, from wasp-like creatures about 120 million years ago in the Mesozoic period. This means that they must long ago have reached 'perfection' along the particular trajectory of their development. They have survived essentially because they are social animals. Although they appear well endowed in terms of sensory organs, navigation and defensive methods of survival in their current environments, it is probable that they will survive only as long as these environments remain favourable.

They have a protective exo-skeleton. The brain consists of a series of ganglia joined at the caudal end and adjacent to two compound eyes. They also have three small ocelli on top of the head that detect light levels and polarised light used in navigation. They have two antennae or feelers attached to the head, which detect chemicals, air currents and vibrations. Two strong mandibles manipulate objects, carry food, build nests and are used for fighting.

Ant colonies vary in size from a few dozen to millions. They are usually built underground and are of sophisticated construction with trails marked by chemicals leading to food sources. This has led to the study of colonies and the application of principles observed in ant colonies have been applied to socio-biological research and computer science. The majority of ants are female worker ants. There are a small number of males (drones) and a variable number of queens. Ants differ in size within a colony. A difference of a single gene determines whether the colony has a single or multiple queens. Female worker ants tend the queen. Fertilisation of the queen happens when she is in flight after which the males that pursue her soon die. Eggs form and develop through the stages of larvae and pupae after which most ants become worker females.

Life in an ant colony is highly complex and organised. They communicate by means of pheromones or chemicals that they sense via the antennae. They leave pheromone trails to find their way about, especially when they leave the nest to find sources of food. A crushed ant emits a chemical, oleic acid, that sends adjacent ants into what has been called an 'attack frenzy'. Dead ants are removed from the nest. Ants attack and defend by biting and injecting or spraying chemicals of various degrees of potency. Nests are often complex but will be quickly abandoned when there is a threat. Some ants are nomadic or form temporary nests. Others build nests by weaving leaves together using silk that they induce their larvae to produce. Most ants are scavengers and indirect herbivores. Additional methods have been devised: some cut and carry small fragments of leaf into the nest and spread these on a fungus, which produces substances that are nutritious. Research continues to reveal new aspects of ant life.

Ant behaviour within a colony is readily describable in terms of human behaviour patterns, which suggests that similarities are easily recognisable. Ant behaviour does not appear pointless or meaningless to human observers, (as like trying to read a foreign language would be). Indeed, it is usual to describe all the activities within a colony in scientific terms, as though the ants have human attributes. The description below is merely a more accurate and therefore more scientific method of doing the same thing. Ants have many methods of foraging for food, gathering food from other species such as aphids and caterpillars, by cooperation, or sometimes by attack with sting or chemical agent. It is evident that they have some degree of reasoning power and they must have memory for reason to perform these functions.

When there are sensory organs and a brain, there must be a diminutive form of consciousness plus memory and reason. A stimulus would produce pain. But in addition, a rudimentary Feeling response of pleasure or some form of opposite plus Reason seems probable. A choice of acceptance or rejection is then possible and action may follow (the sensory-motor response) to accept or reject. Reason as humans understand it produces the possibility of choice. In the ant world, choice would apply when ants need to find their way back to the nest, for example, (or in humans back to their homes); or when there is a choice between fight or flight. It is surprising how a brain as small as a few ganglia is sufficient to achieve the results observed. But if one rejects that this description is the explanation for ant behaviour, then one has to ask what alternative mechanism could explain the facts?

8 Twenty First Century Philosophy

Most people would agree that genes must be in control of all bodily physical structures and functions from the fusion of sperm and ovum that brings an individual into existence until the point of death. Genes are therefore also in control of all Minomic functions; otherwise there is no explanation for the existence of these. We can be sure of both statements because no action by *individuals* can affect these functions, but all are free to apply Abilities and skills and all would accept that these are different for each individual.

Minomic functions, such as mathematical and musical Ability for example, vary greatly from one individual to another. The reason for these variations is unknown, but must be gene based and is under investigation by biological scientists. The variations from individual to individual may be profound and often have far-reaching consequences for societies. It appears probable that these variations were a feature of the Brain Age throughout its evolutionary history. In the 1920s and 30s Konrad Lorenz (1903-1989) [29] of Austria and his colleague Nikolaas Tinbergen[30] observed in detail the behaviour of birds, particularly goslings and jackdaws. He explored the phenomenon of 'imprinting' - the fact that after birth they take the first object they meet to be their mother and become attached to this object and even artificial modifications of it. He regarded these as 'instincts'; innate properties of Brain-Mind function and described 4 components, - 1) they are innate, 2) species specific, 3) behaviour specific and 4) have action specific energies and releasing mechanisms; or in Minomic terminology, the appropriate Abilities and Sensibilities.

In 1978 Premack and Woodruff asked the question 'Does the chimpanzee have a theory of mind?' [31] This was taken to mean 'know', 'want', 'see' 'believe' etc. In other words, 'do these elements of human understanding exist in chimpanzees?' which is, of course, an entirely different question from that investigated by Lorenz in non-primates. Ever since, investigators have focussed on this question.

Professor C M Heyes, an evolutionary psychologist made a detailed survey of the voluminous literature on how animals communicate with one another[32]. Many theories have been proposed and experiments on animal behaviour have been designed to obtain conclusive evidence.

---

[29] *The Companion in the Bird's World*, Konrad Lorenz, Auc 1937, 54, 245-273.

[30] *The Study of Instinct* N Tinbergen, 1954 Oxford, The Clarendon Press.

[31] *Behavioural and Brain Sciences*, 1978, 4 515-526.

[32] *Anecdotes, training, trapping and triangulating: do animals attribute mental states?* C M Heyes, Animal Behaviour, 1993, 46, 177-188.

Do they have beliefs about the world around them: do they have a theory of mind? Heyes suggests that investigators have not appreciated that experiments have not excluded the possibility that communication is by associative learning and not using mental states familiar to humans.

Heyes made an extensive review of the literature in non-human primates[33]. She begins by asking the question: does *any* primate have *any* capacity to conceive of mental states? She reviews theories of mind based on categories of behaviour: imitation, self-recognition, social relationships, deception, role-taking and perspective-taking. She found that there is no convincing evidence of mental states. She proposes an elaborate experiment designed to answer the question decisively[34].

She describes the widely accepted current view of evolutionary psychologists that the mind consists of many pre-existing innate cognitive modules that are mirrored in the brain and thus make possible the evolutionary analysis of cognition by comparisons of brain-mind functions. But she adds that there are now many aspects of the subject that do not fit this relatively narrow framework without the risk of distortion and error. Heyes suggests a comprehensive plan of 'Four Routes' of Cognitive Evolution to embrace all the current research methods of workers in this field. Route 1 is by *Natural Selection*, which makes possible *Phylogenetic construction*. If this should lead to change in the organism, it is by *Inflection*, which is Route 2. Heyes reviews changes that have been reported such as imprinting, spatial memory in birds of where food has been stored, face processing, theory of mind and imitation, but finds the results unconvincing as these could have occurred through one of the two alternative Routes, as follows.

Change brought about during the lifetime of an individual organism, is called *Ontogenetic construction*, Route 3 (these are potential changes to the ontogeny - the being, body, or phenotype - brought about by interaction between the developing organism and its environment). When changes actually occur they are due to *Inflection* of the ontogeny, which is Route 4 (the means whereby the subject influences the outcome). The success of each mechanism must depend upon the adaptive capacity of the organism (the capacity of the organism to utilise or adapt the Abilities that are available), for without these, all such changes would not be possible.

---

[33] *Theory of Mind in Non-Human Primates* C M Heyes, Behavioural Brain Sciences (21 (1): 101-134

[34] *Four Routes of Cognitive Evolution* C M Heyes 2003 Psychological Review, Vol 110, No. 4, 713-727

Heyes concludes her description of the 'Four Routes' by adding: 'What is distinctive about evolutionary analysis is that it investigates not just mature form but origins, and phylogenetic and ontogenic processes are the two major alternative [evolutionary] sources of mature form'. My description of the Minome as a function within consciousness accords well with Heyes description of Developmental Selection, Ontogenic construction and Inflection. Moreover, it shows how the Minomic manifestation of the Evolutionary gene relates to gene expression as a Phylogenetic construction (see further description of this topic in the Conclusion).

How is Heyes' 4 Routes classification to be modified to allow for gene control *within consciousness*? Phylogenic construction (Routes 1 and 2) and ontogenic construction (Routes 3 and 4) now become possible simultaneously. Ontogenic constructions trace the pattern of each person's daily life. As explained, these determine the fate of individuals and of the leaders in society that some of them may become. But since there is also phylogenetic gene control (Route 1), what possible inflections (Route 2) could occur? From within consciousness, all subjects are clearly aware of the limits to Ability levels imposed by extra-Minomic gene control (some have sporting Abilities superior to others; some musical Abilities, or the Ability to do research not possessed by others). It has to be kept in mind that the effect of Gene-Consciousness (see p 31) is to maintain ontogenic constructions from generation to generation. They therefore inevitably have a continuously evolving content that is controlled by Reason, the Sensibilities and Conceptual Ideas.

Is there evidence of phylogenetic inflection operating outside the Minome? This would raise, lower or otherwise change Ability levels available subjectively within the Minome. These would become evident over millennia but could become evident within a single generation. These possibilities are discussed in the Conclusion under Gene-Consciousness. There can be little doubt that the addition of Gene-Consciousness to memory must have added enormously to the effectiveness of memory in Brain Age animals, but in particular humans.

What is the significance for lower animals of these changes, which must have operated throughout the Brain Age? We have to consider in particular the three Sensibilities and how these could have changed throughout evolution. Feelings dominate throughout the Brain Age as they still do in humans. This was inevitable as the design of the animal kingdom is built on the basis that animal attacks, kills and eats animal. And bearing in mind the emphasis on cuisine involving animals today, the effect is to promote the same evolutionary design. Feelings are modulated by Aesthetics and Values. Evolutionary psychologists look for Aesthetics in birdcalls for example. But Values are largely confined to humans. It follows that throughout history and right up to the 20th century of war and disaster, this Sensibility must be regarded as weak. All of this work, however, is still in its infancy.

In conclusion therefore, since the beginning of the Brain Age, all ontogenic activity is assumed to have been within consciousness. The activities of the Minome with its sensory-motor extensions covers the territory from which science obtains its data. But science, focussed on the objective universe including the brain, will not solve evolutionary problems. This is because animals, like H Sapiens are exposed to the Sensibilities plus Reason and Conceptual Ideas and it is from these that understanding is derived.

Returning to Heyes' review of the literature of mental capacities in non-human primates she begins by asking the question that Premack and Woodruff [35] asked 20 years ago: 'do any non-human primates have the capacity to conceive of mental states?' The conclusion she reaches is that there is still no clear evidence in support of the idea. That really means there is only limited understanding by primates in the sense that humans 'understand' in Minomic terms. Baboons, for example, are capable of deception, as shown by the incident quoted by Heyes in a personal communication to her: 'a male, who does not willingly share, caught an antelope. A female edged up to him and groomed him until he lolled back under her attentions. She then snatched the antelope carcass and ran'. The key to discovering why the baboon ran off with the antelope carcass is to be found by investigating the Sensibilities and not in reason as it is applied in scientific discovery. The Feeling Sensibility is complex. The baboon was showing evidence of being able to manipulate this Sensibility using Reason and appropriate Conceptual Ideas.

Clearly in retrospect Premack and Woodruff, had asked the wrong question. The baboon was using simplified Minomic functions. No attempt has yet been made to investigate how the human Minome has to be utilised to understand lower order animals (in the manner applied in the case of ants as described above), and again it should be noted that a 'theory of mind' using the Minomic system cannot be achieved by reason alone applied to the sensory experiences of non human primates.

Heyes reviews the evidence that chimps are like young children but concludes that they are *not* like children. To answer Premack and Woodruff's question she says we need more strong experiments, not more weak arguments.

---

[35] *Does The Chimpanzee have a Theory Of Mind?* Premack, D and Woodruff, G, Behavioural and Brain Sciences, 1978, 4:515-526

But she does provide a 'strong argument' herself in the same article when she quotes the suggestion by D. C. Dennett that: 'field ethnologists observing animals in their natural environments, the intentional stance is easier to use than the languages of behaviourism or information processing' [36]. (Here he uses the term 'intentionality', originality proposed by the mathematician Husserl, as explained in BHVS).

In Minomic terminology, the 'intentional stance' is equivalent to motivation (the motor side of the sensory-motor axis), which has been a fundamental property of life from unicellular organisms to H. Sapiens, but of course movement is equally a property of the inanimate world of quantum physics, for energy drives the universe. We use 'motivation' or equivalent terms to explain the dynamism of the animate world because it is that which makes possible the sense of understanding or meaning to all phenomena. But what is not generally allowed for by philosophers, but was by Hume, is that the cause of all flux in the universe is unknowable and this applies both to the objective phenomena of quantum physics as well as the subjective phenomena of the Sensibilities. Only when these two dimensions of the unknowable are built into philosophy will descriptions become more accurate (the scientific method allows scientists to ignore unknowables). The subject of philosophy would probably not exist, certainly not in its present form, if it were not for these unknowable phenomena! Let us see what happens when an attempt is made to abolish them.

In his book 'Freedom Evolves'[37] Dennett describes a materialist/determinist universe (in which by definition unknowns do not exist). His objective is to show that the consequences of materialism do not bring about an inability to choose. Although he shows that choices are possible, nevertheless he does not appear to accept that the ethical veracity of such conclusions is flawed. Arguments about the consequences of materialism raged a century and more ago. At that time equally problematic was the relation between reason and ethics, which spawned pragmatism. But moving between materialism and ethics Dennett does not find a problem, yet the unknowability of the origin of the Sensibilities is an integral component of ethics.

If one does not accept these solutions, what are the alternatives? A vast amount of work is in progress.

---

[36] 1983 Intentional Systems in Cognitive Ethnology, The 'Panglossian paradigm' defended. *Behavioral and Brain Sciences* 6:343-390. *Intentional Behaviour,* editors: A C Montefiore and D Noble. Published by Unwin Hyman.

[37] *Freedom Evolves*, 2003 Daniel C Dennett published by Penguin Books 2004.

W. C. Winsatt[38] says 'Attempts to find biology in culture seems either to push towards sociobiological conclusions drawing connections between genetic and cultural varieties and practices, or towards a perception of cultural change as somehow analogous to biological evolutionary processes'. He wishes to find a relation between, 'both individuals and transmissible units of culture (between ideas, practices and pieces of material and software terminology)' on the one hand and biological genetic units on the other. In the Darwinian context, how are the fittest who strive for survival to be defined? If the multifaceted description of culture given above within the Minome is correct, then there seems to be no problem since the 'cultural units' are already within the gene based Minome.

But scientific understanding of brain-mind relations continues to evolve and in this context, Heyes discusses the 'mirror neurone system'[39] and reviews the relevant literature in the human and the monkey. Scanning devices applied to the premotor and parietal cortical areas of human and monkey brains reveal what have become known as 'mirror neurones'. These light up in response to matching signals from an agent. The evidence shows that the mirror system in the human is neither wholly innate, nor fixed once acquired. As Heyes suggests, the mirror system is both a product and a process of social interaction and is dependent on sensorimotor experiences. Heyes et al[40] conclude that 'it contributes via its roles in language acquisition and theory of mind to our capacity for complex social interaction and also depends for its development on the availability of correlated sensorimotor experience in the sociocultural environment'. The implications of these findings are discussed below.

---

[38] *Genes, Memes and Cultural Heredity,* W. C. Winsatt, 1999 Biology and Philosophy14: 279-310
[39] *Where do mirror neurones come from*, Heyes C, Neuroscience and Biobehavioral Reviews. (2009), doi: 10. 1016
[40] Sensorimotor Learning Configures the Human Mirror System, Catmur C, Walsh V and Heyes C, Current Biology 2007, 17, 1527-1531.

9 Conclusion

Conceptual Ideas appear in consciousness that may become the precursors of beliefs and knowledge. When this happens, they have motor potential, that is they determine objectives within the Minome. Is there evidence about how objectives are formed? Both Leibnitz and Kant were aware of *apperception*, the Ability of two or more streams of Conceptual Ideas to operate simultaneously. This must be true, for one neurone system, the brain, can be trained to drive a car and at the same time the driver may engage in a vigorous conversation with a passenger. Also, the mind is able rapidly to review alternative objectives, which is the essence of thinking and making choices. But choices are influenced by numerous other peripheral factors such as possible risks, financial problems, and the tooth that has been painful for a few days, etc. The Ability to think at various levels makes possible the rapid review of alternative choices. The work of Professor Heyes and others is beginning to reveal systems of cells that make these apperceptual functions possible.

Motivation prior to the Brain Age was a gene-based activity that energised all life. The nature of this driving force is unknown but prior to the Brain Age it produced millions of species, most not being fit enough to survive. *Always the objective of the evolutionary process seems to have been to produce more effective and efficient ways of reproducing life.* It has to be emphasised that the gene-based Abilities within the Minome available to humans determine the direction of our endeavours. But there is no evidence of a final genetic destination, such as Aristotle postulated. Brain Age animals came to dominate the living world presumably because the brain is more efficient and effective than the random creation of new species by the gene. But the elements of understanding have become increasingly complex during evolution and in H. sapiens they take the form of objectives to be achieved within the Minome. This has had a profound effect because the main objectives now are not the production of new species by the gene outside the Minome, but the search for *human objectives by each individual within the Minome* (greater happiness and protection against all the threats to human existence, for example).

In summary, genes operating outside the Minome determine the extent and degree of all mental and physical Abilities for each individual, but the modified function of these genes also determine Minomic functions *within consciousness* that give the sense of 'freedom' to make choices, which is the beginning of understanding. The gene therefore oversees everything that happens in consciousness, but operates through Minomic functions, so that its manifestations are different from the usual phenotypic evidence of gene activity *outside* of the Minome.

Doubtless it will be argued that the Minomic structure described excludes the possibility of real freedom. But to propose an indefinable something called 'freedom' and then seek its meaning, is to engage in a metaphysical discussion that is meaningless and endless.

Gene-Consciousness

We are so used to life within the Minomic framework, that many, even after a scientific training, do not appreciate the full significance of the scientific revolution produced by the Greeks. This alone caused an enormous increase in size, efficiency and effectiveness of Gene-Consciousness and this continued at an accelerating rate after the Renaissance. This discovery made it possible to employ reason in two quite different but equally potent ways. Reason directed outwards to sensory experience of the inanimate universe and all animate beings within it, I call the α Reasoning of science. All other reasoning is Minomic and normative, and this I call β Reasoning, which is the *response* of the Sensibilities to all sensory experiences.

It is assumed here that gene control continued throughout the Brain Age, which means it is a property of all intra-Minomic *and* extra-Minomic functions. It appears probable that extra Minomic functions dominated at the beginning of the Brain Age and as explained, intra-minomic functions dominate in humans at the other end of Brain-Age development. Since the energy and motivation of all Minomic functions is gene-based, then the end result of what the Minome does is in essence to determine where motivation is directed. That is to say, 'we', using the tools available to us in consciousness (in essence β Reason plus the Sensibilities) decide the objectives. And these differ according to where and when we happen to live in the world. In other words, since the Sensibilities determine the objectives, then our Conceptual Beliefs (which are from the motor side of the sensory-motor axis) determine the directions in which our motivations are channelled. But these will only have significance in a society when the majority agree on a specific objective. And the 'majority' in democracies includes individuals possessing a wide variety of belief systems. Choices in this context mean what they appear to mean, and are certainly not amenable to solutions arrived at by α Reasoning alone.

This is the mechanism by which progress within the Minome was brought about throughout the Brain Age. The objectives were always restricted however, by the trajectory of each individual species. Evolution that is Minome driven becomes permanent only in so far as communities so decide. There is little evidence at present that these objectives became gene based by 'absorption', but recent work on the human genome will soon produce additions to current knowledge about such matters.

The limitations that apply to social activity are due to the physical and mental Minomic limits applicable to each individual in a world limited also by their specific sensory experiences. The content of Consciousness, as defined by the Minome, obviously varies hugely from individual to individual. One may think of gene-Minome relations in various ways. For some, Conceptual Ideas in consciousness may be thought to replace the functions of the gene. For others the impression may be gained by some individuals that they are independent agents free to 'move around in consciousness' because they are able to choose and possibly achieve whatever objective they desire, which is what Lamark and Baldwin claimed. But philosophical Conceptual Beliefs or theories are always in danger of being proved wrong in the sense that scientific theories may prove to be wrong. *But it will never be possible to deny the veracity of Minomic functions per se for these are not theories and exist, albeit with considerable variations, within the conscious experience of every individual.*

I call gene activity within the Minome the *Gene-Consciousness* of Brain Age animals, but the evolving Minome may develop faults that cause an individual's psychophysical behaviour to depart from the norms of society. Thus, experience suggests that an individual who commits murder or serious bodily harm is likely to possess a range of Sensibilities that is confined to little more than the Feeling Sensibilities. Such people may have little or no awareness of Values and may even assume that strong Feelings may equate with what others mean by Values. Others may have a minimal awareness of values but are well aware of the norms of society, yet nevertheless motivation from Feeling Sensibilities easily overrides their sense of Values. Also, the precise sexual orientation of these people is crucial and determines their objective, yet other members of society may be completely unaware of it. Progress will depend on the investigation of the Gene-Consciousness of such people. Although psychologists are working on these problems, especially in children, progress will depend on much more research into the state of the Minome in normal and abnormal subjects.

The position at present is that within Gene-Consciousness, we have the freedom to modify all activities within defined limits. These limits are determined by genetic inheritance at conception. They are the physical and mental Abilities available and the outcome in practical terms depends on the physical and mental conditions that exist for each individual. There are anomalies of brain-mind function that are amenable to modification by appropriate therapy, such as the Psychoneuroses. Others are not. Thus a homosexual can modulate behaviour, but with or without 'therapy' cannot become a heterosexual. Only genetic manipulation could theoretically make such a difference. The cause of more severe brain-mind disorders such as the Psychoses is insufficiently understood at present.

Progress now seems to depend upon achievements within the Minome (such as attempts to understand the universe, human values and the ordering of societies). Gene mutations have not produced new Abilities or a rapid generational *enhancement* of one or more Abilities that exist at present. Instead, what is observed are all the *syndromes of malfunction* known to medicine. Also observed are cancers that affect virtually every tissue in the organism. But apart from such anomalies this strongly suggests that within the Minome, particularly the mature Minome of H. sapiens, human objectives driven by the Abilities of the sensory-motor axis are always strong enough to prevent, or perhaps take the place of genetic-mutations that would result in trajectories superior to those that exist today. The move in societies to accept agreed global objectives for many human societies seems likely to strengthen this position.

Theory of Mind Implications

Since Conceptual Ideas plus Reason enter the mind in a continuous stream, new ideas appear from moment to moment in rapid succession. Many of these become beliefs about the world of sensory experience that include the community in which the individual lives. This enables individuals to decide what to do immediately, within the hour, or in the next day or two. Without this forward planning information, human existence would not be possible. The more permanent extension of this facility in the form of knowledge of the sensory universe enables humans to gain a vision of life far into the future. This makes possible modifications to forward planning and choices due to technology based on this information. These Abilities were obviously extremely limited at the beginning of the Brain Age, but must have slowly extended as more complex animals evolved. It would be possible to describe a prospective history for each animal studied.

The converse of this use of sensory experience is the familiar retrospective history used to understand the past. This sensory experience is essential of course to the scientist when forming an hypothesis.

The sub-atomic dimension also forms a fundamental component of knowledge. Here is the source of the energy that drives the universe and all life within it, including all gene control of Brain-Mind relations.

I mention these dimensions of knowledge in order to consider the relationship of these to current objectives in animal biology, which at present are primarily focussed on animals most closely related to humans, such as the great apes with brain weight approximately one half the weight of the human brain. Little was known about their cognitive abilities. But in about 1960 Jane Goodall, studying chimpanzees discovered that they could make and use simple tools of wood and stone. Then the question became: 'do such animals possess any of the cognitive abilities of humans'? In 1978 this turned into a search for the answer to the question of Premack and Woodruff: 'does the chimpanzee have a theory of mind?' with negative results so far, as described above.

But consider the *prospective history* of Gene-Consciousness in humans and the higher apes. The apes are on a trajectory that may well have reached its zenith. It is difficult to understand how their dependence on trees can lead to future progress. More probable is their extinction at the hands of humans who destroy the trees on which they depend. Humans are on a much more advanced trajectory with no clear end point and are travelling rapidly away from that of chimpanzees. Thus, if positive answers to Premack and Woodruff's question were to be found, it cannot have prospective historical significance, for the two species are like ships that pass in the night, never to meet again.

How are we to approach the problem of understanding in animals? We can be sure that Values Sensibilities are not a part of their understanding so that anything resembling an idea is never faith-based. Also, ideas will not be in the form we know them in H.sapiens, for ours are language based. There is some evidence for the existence of Aesthetic Sensibilities in birds but this is miniscule by comparison with Feelings. This leaves the Feeling Sensibility as the major medium for communication and action in animals.

When Feelings are described in words, they appear as an ill-defined jumble of ideas of indefinite number, such as desire, conviction, happiness, love, hate, cunning, good, evil, anger, misery, despair, anxiety, confidence, etc. However, Feelings are not expressed in this elemental form but in a more accessible manner. Since Feelings are responses to sensory experiences, many are oscillations between visible limits; such as good - evil, happiness - misery, love - hate, etc. If there were no oscillations, how could they be recognised? They could be mistaken for sensations that do not change. But some Feelings such as anger are different in that they simply appear and after a short duration, fade away. Humans are also able to remember these events and use them on future occasions to achieve desired objectives.

Gender differences were built into the genetic and phenotypic structures of most species from the earliest pre Brain Age animals. Departures from the norms, such as sado-masochism and paedophilia have long been rejected by communities. But are comparable problems ever observed in animal communities?

The number of distinct Feelings is uncertain because literati are continuously inventing new expressions of feeling. This wealth of the Feeling Sensibilities is the material humans draw upon to describe the evolutionary process. Understanding the evolution of Brain Age animals requires descriptions in minomic terms of the sequence of events in a large variety of animals from the early times up to the present. It is emphasised, however, that the biological evidence at present is extremely limited.

One problem that presents in the early Brain Age, at the level of the ant, for example, is whether gene activity *outside* the Minome dominated, and the Minomic contribution was small and insignificant, or whether, since the ant has sensory organs, brain plus mind *within* the Minome produced the phenotypic result observed, as in more advanced animals? If outside, the ant's complex behaviour in colonies could have evolved by trial and error via many species of ants known to have evolved but failed to survive until a species appeared that was capable of survival. If this seems unlikely, then it must be assumed that ant colony life is the product of a Minome that is surprisingly mature, considering the minute brain.

Another phenomenon must also be mentioned before proceeding to describe animal understanding. This is the fact that in the process of describing evolutionary changes, humans are discovering new ways of characterising the unknowable, but are not diminishing the barrier if the following is true. Surveying the Brain Age from its beginning to the present day, reveals enormous strides in the scale of understanding that occurred over millions of years, yet there is no evidence that at any stage animals played other than a passive, responsive role. An unknown mechanism produced continuous changes in Gene Consciousness to achieve the result in humans described above. This included the appearance of the Aesthetic and Value Sensibilities some thousands of years ago. It remains a fact that humans are still only able to respond to what Gene Consciousness allows them to do. In essence, this is what Aristotle was searching for in his Metaphysica [41] and I have therefore called it the Gene Mover (after his Prime Mover).

We judge the animal mind in terms of what we become aware of as a result of our own observational experiences that are assumed to have some comparability with those of the animal. That is what Jane Goodall did in her study of chimpanzees.

---

[41] *The Works of Aristotle, Volume VIII,* Oxford at the Clarendon Press 2nd Edition 1928.

This is potentially hazardous, but unavoidable for anyone working in the field. Therefore this position has to be accepted as a convenient base from which to take the next step towards describing animal understanding by translating what is observed into Minomic terms. All the elements of the Minome listed on page 7 except for the Aesthetic and Value Sensibilities are active but their contributions differ from those in the human and therefore they will not be discussed in the order listed.

I begin with Perception. The external world is revealed by vision, hearing and other senses. These determine in large measure the world in which the chimp lives. To what extent does the chimp's knowledge extend beyond its immediate environment? This will depend upon perception, the continuous flow of Conceptual ideas, Reason and a Memory that makes it possible to utilise these components of the Minome. These determine, for example, whether the chimp's world is confined to the valley in which it was born and if not, how far does it extend?

Social life depends on Perception, the Conceptual ideas that motivate Abilities for the relevant Feeling Sensibilities, α Reasoning and α Memory. The Conceptual ideas create the world in which the animal lives. The Sensibilities will vary in variety and intensity over time and from animal to animal.

Goodall described aggression, killing and cannibalism. Clearly, the Conceptual ideas formed, the Abilities generated and the relevant Feeling Sensibilities, Reasoning and Memory are going to be different.

It may be asked: What is the advantage of converting to a Minomic terminology? If we are to pursue advances in Brain-Mind understanding, then we have to use the elements of understanding set out in Minomic terms because these describe causes rather than their effects expressed in arbitrary linguistic terms. Minomic terms provide a standardised algebraic system of symbols that stand for causes. It may prove useful to grade each of these components on a scale of say 1-5, when more observations have been made in variety of animals.

The Evolution of Philosophical and Psychological Ideas

The crucial change that transformed Consciousness was the invention of the Greeks that brought what came to be called 'science' and 'philosophy' into existence. But that is not how it began, for 'philosophers' and 'scientists' began much earlier to use and apply reason to seek truth and falsity in Babylonia.

Babylonian civilisation lasted through many dynasties from over 3 millennia BC to Greek times. During the Old Babylonian period of about 15 centuries, astronomy evolved from older faith-based beliefs to become science based and dependent upon reason, logic and mathematics. Literature and architecture became established branches of knowledge. Medicine also developed to include many Hippocratic features. Later, civilisation from the 7th century BC to Greek times included the further development of astronomy, mathematics and the appearance of elements of philosophy, including speculation about sensory experience in general. This led Seleucus (b 190 BC) for example, to propose the heliocentric theory that the earth rotates on itself and also rotates around the sun. Copernicus and Galileo confirmed these results almost 2 millennia later, early in the Renaissance. Most discoveries in mathematics, however, were made during the Old Babylonian period.

Early philosophy often took the form of sayings and stories that served to teach the ethical norms of society and tied this to beliefs about the gods, all of which was in line with ancient eastern religious beliefs. This was expressed in the form of dialogues, epic poetry and folklore.

In Greek science, Heraclitus, (about 500 BC) taught only what could be observed, namely that everything *observable* is in a state of flux. But Parmenides (in poetic form) and Zeno of Elea *postulated* that there must also be a permanent, fixed Reality. Democritus *postulated* that this immutable element is the atom (that which has no parts) and all observable movement is due to the fact that atoms continuously change position with respect to each other. But at this time faith based beliefs were still strong. For example, in 5th century Greece, the Olympian Gods and the Delphic oracle held sway, which led to the death of Socrates because his ideas infringed beliefs. But the extent to which this was motivated by faith-based ideas or was of political origin is uncertain

By the 4th century, Platonic transcendentalism and Aristotelian teleology made the Olympians irrelevant for many of the elite although not for the 'foot soldiers' that tended to adhere to the old beliefs. Plato (428-347 BC) taught by means of the Socratic Dialogues. The search for knowledge and understanding was as alive then as it is today. In the Theaetetus these were discussed but no conclusion was reached.

Plato's own view was that the sensory world we are aware of is transient and of little significance. The real world we are only faintly aware of as 'Forms'. Evil in the sensory world is due to lack of knowledge. Only prolonged training of suitable individuals both in what is good, as well as in philosophy (science) and mathematics will discover goodness and that, once discovered, will never be lost. The ideal society exists independently of humans but has to be discovered and those who fail to discover it have to be ruled by those who do. Platonism does not produce a free society that is the objective today because it does not rest on discovering the brain-mind psychology of each individual, which is where the focus of attention now lies.

Aristotle brought Plato's transcendent world down to earth, teaching that what is real is the visible Forms of objects, which are in continuous motion. This describes their state at any given moment. He arrived at this position after reviewing the conclusions of Greek philosophers up to the time of his contemporaries, Anaxagoras and Empedocles. He does this in Book 1 of Volume 8, the Metaphysica.

He summarizes his position as follows. The first sentence of Book A1 (980a) states: 'All men by nature desire to know'. In the 2nd paragraph, 'By nature animals are born with the faculty of *sensation*, and from sensation *memory* is produced in some of them'. Together these make possible the use of experience. Multiplicity of experiences makes *wisdom* possible. Master workers in each craft know more and are wiser because they have knowledge of causes. Manual workers perform tasks routinely but do not know the causes. In the 3rd paragraph he summarizes several important points.

1. Multiple observations of a specific sensory experience (α Reasoning) form the essence of science. These individuals can teach, discover new things and are wise. They are the true scientists of yesterday and today. This method tells us the 'why' or cause. But he emphasises that scientific knowledge per se is useless. To acquire utility, the following is necessary.

2. Multiple observations of communal activities, which Aristotle calls 'art' (β Reasoning: Reasoning plus Responsive knowledge from the Sensibilities). These give pleasure and the necessities of life but the 'why' or cause comes from science.

The physician, he states is in a unique position for he utilises knowledge from 1 and the art from 2. Aristotle explains that 'The reason for this is that experience is knowledge of individuals, art of universals; and actions and productions are all concerned with the individual; for the physician does not cure *man*, except in an incidental way'. He treats an individual person. But today the 'incidental way' is in many respects paramount as preventive measures including genetic manipulation are becoming of increasing importance.

In the fourth paragraph of Book AI of the Metaphysica he discusses in more detail the α Reasoning of point 1 above and the β Reasoning of point 2 in relation to *knowledge* and *understanding*. Wisdom requires both of these. He concludes: 'clearly then Wisdom is knowledge about certain principles and causes'. He then embarks on the search for first principles and causes that must explain the Forms and is the content of the rest of the Metaphysica.

Before arriving at that point, however, Aristotle analyses in Volume 3, Book 1 (338a) of the Meteorologica, the very question that is examined in this book, ('A New Approach to Evolution'). He does this in a very rigorous, logical and scientific manner. His aim was to answer the questions: does the soul exist? How is this to be proved and what are its properties? He searches for a motivating force that could account for motivation, for the activity of all animate beings. He reviews the opinions of all pre Socratic and contemporary philosophers. He notes that all mythology produced faith based answers in the form of gods and divine heavenly bodies that are stationary or move in circles. In Book 11 (412a) he sets all these theories aside and embarks on his own explanation. The concept of Forms he had justified by defining it carefully with the aid of such terms as essence, substance and matter. It would take us too far afield and is unnecessary to follow his arguments in detail as it turns out that his conclusions are quite simply summarized as follows.

    He begins his analysis of the existence of the soul by examining the problem of movement. All Brain Age animals have a sense of touch and the most advanced have eyes. Nutrition during life ensures growth and all necessary activities of body and mind to achieve this, which implies movement. The highest activity of mind is thought. In old age, it is not the mind that declines but the body. 'The incapacity of old age is due to an affection not of the soul but of its vehicle, as occurs in drunkenness and disease. Thus it is that in old age the activity of mind or intellectual apprehension declines only through the decay of some other inward part; mind itself is impassable' [42]. He concludes that only a soul could sustain the mind in life and after death.

    The rest of his proof is straightforward. Aristotle takes all the elements of understanding listed above on page 7 and attributes these to properties of the soul. The alternative is to regard them as simply unknowable. Aristotle maintains that the concept of the soul is necessary to sustain these activities during life and that logically in order to do this, it must itself be immovable. Therefore the soul must have properties of the One, and in the Metaphysica it is identified with the One, that is, with God [43].

---

[42] *The works of Aristotle Volume 3,* Book 1, 408b 20
[43] *The works of Aristotle* Volume 8 1072b, 0-39

In Hebrew literature, laws had to be obeyed because they were God's Laws, but in 4th century Greece, individuals sought solutions within societies. Later in the century Epicurus, an atomist, taught the ethics of Psychological Hedonism, but added that some pleasures are good and others bad and the latter should be avoided. His science and ethics were not logically related[44]. Zeno of Citium, taught Stoicism, (from stoa, a porch), a largely ethical system that evolved through many versions in the Greco-Roman world. Stoicism was largely a response to the depressive effects of social turmoil after the collapse of the Greek city-states and the Alexandrian Empire. He taught that one should learn to become indifferent to adverse events.

In Greek civilisation, what is now called psychology had not emerged as a discipline. But its content was present in the minds for example of leaders in society of religious, political and military organisations. Philosophers sought to understand the origins and future of all these elements.

The Enlightenment was a complex period of philosophical investigation dominated by the belief that in time reason would solve all philosophical problems. The Romantic Age undermined these endeavours and no satisfactory solution to the divide between science and ethics was achieved. The Scholastics had described the immortal part as the soul. Descartes pointed to the dualism involved and that mind could not be extended in space. The Victorians put it this way: 'what is matter? - never mind. what is mind - no matter'.

The Gene Mover

Returning to the concept of the Gene Mover, whatever its nature turns out to be, it will not be the unmoved mover of Aristotle. In view of the evidence from evolution, this agent appears to have directed motivation of the Abilities towards *maximizing the possibilities for life on earth.*

Gene Consciousness allows freedom in the sense of choices that are clearly greatly superior in humans to those of any other species. But the choices are always within the limitations imposed by the Gene Mover.

Today the problem takes the form of the psychology of brain-mind relations, discussed here from the evolutionary standpoint. A new approach to the problems of understating and knowledge acquisition is described in this volume. The current methods were found to be wrong in many respects.

---

[44] See BHVS, discussion 4, for a more extended and relevant description of this period.

10 Glossary

Analysis	The investigation of concepts by linguistic and logical means.
Apperception	The Ability that enables two or more streams of Conceptual Ideas to operate simultaneously.
Aristotle	Volume 8, The Metaphysica
Associationism	18th and 19th century theory of ideas and how they are formed.
Behaviourism	A series of doctrines designed by psychologists to ensure that psychological descriptions accord with those of science.
Belief, in science	A tentative truth: a belief or hypothesis in science that has been shown on factual evidence to be 'true'.
Beliefs and…	'Beliefs and Human Values' (BHVS). Book by Richard Sheriff Jones, Trafford Publishing, 2009.
Belief, normative	A moral truth or principle accepted by an individual or community as guidance.
Belief, faith-based	No evidence is offered or is possible for a belief or system of beliefs.
Clade	A group of organisms that have evolved from a common ancestor.
Deconstruction	Derrida in the 1960s showed that any system of ideas could be 'deconstructed' and thereby shown to be false.
Disposition	Tendency to behave in a certain way.
Dualism	The theory that mind and matter are two distinct things, as asserted by Descartes - mind cannot consist of spatially extended matter. It must therefore be non-material, which is soul.
Empirical	From experience, not theory.
Epigenetics	Changes in phenotype expression (appearance) caused by mechanisms other than the DNA sequence.
Epistasis	Gene-gene interaction.
Ethology	The science of animal behaviour, or comparative study of animals.
Foundationalism	The theory that knowledge of the world rests on a set of indubitable beliefs.
Functionalism	The theory that a conceptual belief is brought about by the functional role it is to perform and not its intrinsic properties or character.
Gene-Consciousness	Gene function within consciousness.

Gene Mover	'Mover', from the Greek of Heraclitus and Aristotle to describe all movement.
Holism	The idea that items such as beliefs, properties, language, etc has a common root.
Instrumentalism	Scientific theories do not represent an unknowable reality, but are tools with which to understand the observable world.
Knowledge	Collection of related beliefs.
Metaphysics	Philosophical theories and explanations (e.g. substance, energy, space, time) that lie outside the scope of scientific investigation using its current methodology.
Minome	The term used to describe concept formation and all aspects of understanding.
Naturalism	(Ox companion to Phil): Epistemology- How we can gain knowledge of the world.
Norms	Rules for the evaluation of conceptual beliefs.
ethical	Ethical norms of society. Excludes faith based ethics.
Observations	Sensory experience that only becomes significant or acquires meaning when theory is applied.
Ontology	Nature of Being; of Existence; of Reality.
Positivism	A term introduced by Auguste Comte and Saint Simon in the mid 19th Cent. Describes the history of human thought: Religious, Metaphysical and then Scientific. Later included Marx and then Logical Positivism in the 20th Century.
Postmodernism	After the modern era. Late 20th century and 21st century philosophy.
Pragmatism	What works in theory and practice is as near to truth as we can get.
Predicate	A subject-predicate sentence in which the subject phrase refers to something or someone.
Realism	The idea that things observed are representations of reality; also called Representationalism.
Reductionism	The theory that the psychological properties of mind can be reduced to physical counterparts.
Relativism	Protagoras, 'man is the measure of all things'. But there is limitless diversity: cultural, racial, and linguistic.
Religion	Faith-based Beliefs.
Representationalism	'Anything that represents something' only acquires meaning because commonly used in linguistic and other communications.

Sensory- motor axis	The sequence of events upon which all life is based: sensory experience; the response to it and the actions that follow.
Synthetic-a priori	Kant's term relating evidence from sensory experience with logical truths .
Transcendentalism	That which is postulated to exist beyond sensory experience. Therefore metaphysical.
Understanding	Rationality, (logical/mathematical reasoning). The Sensibilities - Awareness of Feelings, Aesthetics and Values.

11 Index

Abilities, 7, 16, 17, 18, 33-34.
Aggression, 35,36
Analytical philosophy and the limits of understanding, 5.
Animals and conflict, 27.
Ant, 14,16,21,23-24.
Ant brain and consciousness, 36.
Apes, African, 18,34.
Apperception, 39.
Aristotle, 39.
Baboons, understanding, 28.
Babylonia, 38.
Bateson, P, Behavioural Biology, 14.
Bees, 22.
Beliefs, categories, 17.
Beliefs or hypotheses of science, 7, 9, 11,18-20.
Beliefs, faith-based, 19.
Beliefs and Human Values, BHVS, 5. Other references to BHVS,10, 15,29.
Big Bang, 6.
Bonner J T and multicellulariity, 14.
Brain and sensory organs, 6, 7.
Brain Age and evolution, 13, 30.
Brain Mind relations, 35.
Brain Size and sensory organs, 18.
Brain Size and Minomic functions, 8-9.
Children (and Chimps), 28,32.
Chimpanzee, 25,35.
Choices, 31,32.
Concept formation, 7.
Conceptual beliefs, 21,32-33.
Conceptual Ideas, 4,17,33,36.
Consciousness, emergence of, 7,11,14, 16,31,32.
Copernicus, 38.
Darwin, 6,13,14,19,30.
Delphic oracle, 38.
Democritus, 38.
Dennett D C, 9,10,11,29.
DNA, 12.
Descartes, 41.
Einstein and scientific knowledge, 12.
Enlightenment, 41.
Epicurus, 41.
Epigenetic Mechanisms, 15,16.
Ethology, 14.
Evolution, 4,5,7,9,11,13,17,20,31, 32,37.
Evolution of philosophical and Psychological ideas, 37.
Faith based ideas, 9,19,35,38,40.
Feelings,7,27,33,35,36.
Forms, 38,40.
Freedom Evolves, Dennett, 9-11,29.
Freedom and choice, 31.
Galileo, 38.
Gender differences, 35.
Gene Consciousness, 27,32,33,35.
Gene Consciousness and the Greeks, 36.
Gene Mover and Aristotle's Prime Mover,36,40.
Genes and phenotypes, 13.
Gene, 'wild', 13.
Gene, within the Minome, 31-36.
Gene, outside the Minome, 31.
Genetic assimilation, 21.
Genome, 14,33.
God, 10,40.
Goodall J, 35,36,37.
Gould S J and Levontin R C, understanding, 19.

Gottlieb G, 15.
Greeks, 32,37.
Griffiths and Neumann-Held, 13.
Hebrew literature, 41.
Heraclitus, 37,38.
Heyes C M, evolutionary psychologist, 25,26,27,28,30,31.
Hippocrates and medicine, 38.
History, Prospective and retrospective, 28,33,34.
Homosexual, 33.
H.sapiens and the gene, 18.
Ideas,   7, 8,10,11,18,37,38.
Innate and Acquired ideas, 14,25,26,30.
Kant I, 31.
Knoll A H, earliest forms of life, 9.
Knowledge, 5, 6, 18,37, 40.
Knowledge and the Gene Mover 41.
Knowledge, four kinds of, 17.
Learning, 18,19,21.
Leibnitz, 31.
Lewis, E B, 6.
Lorenz K, Bird behaviour, 24.
Introspection, 2.
Materialism and choice, 28.
Mathematics, definition, 5,37.
Mameli, M, 9,17,19.
Meaning, 2,4.
Medicine, 34,38.
Memory, 7,20,24,27,7,39.
Metaphysica of Aristotle, 39,40.
Metaphysics and terminology, 20.
   and review of causes, 30-31.
Mind, 10,15,20,25,30,33,34,37,38.
Minome, 5,9,18,35,36.
Minome, pre and post Brain Age, 7-8, 11,7,17,20-22,27,30-34.
Minome, gene control of, 15,33.
Minome and Evolution, 21,34.

Minome and theory of mind, 27.
Minomic terminology, advantages, 36-37.
Mirror neurone system, 30.
Molecular gene, 13,14.
Motivation, 17,29,31,33,40,41.
Murder, 33.
Mythology, 40.
Naturalism, Quote from Dennett 9.
Neanderthal Man, 18.
Old age, 40.
Olympian gods, 38.
Paedophilia, 36.
Perception, 7,37.
Perfection, principle of , 16.
Perfection of gene products, 16,17,23.
Parmenides, 38.
Personality, use of unsatisfactory
   terminology, 12.
Phenotype, 13-14,26.
Philosophy, origins, 37.
Philosopher, role of, 9.
Philosophy defined, 4.
Philosophy and the Mirror of Nature, 10.
Philosophy, boundaries of, 11.
Philosophy in the 21st century, 25,29.
Philosophy, and evolution, 37-38.
Philosophy, Platonic, 38.
Physicalism and Mameli, 12.
Physician, 39.
Plato, 38.
Pragmatism, 29
Premack D and Woodruff G, 25,28,35
Primate cognition, Byrne R W, 18
Primates, 25,28
Prophet, Old Testament, reference
   to the ant, 22
Prospective history, 34,35
Psychology, 11,20-21
   and In animals, 39,41

Psychoneurosis, 33.
Psychosis, 33.
Reason, 5,6,8, 17-24.
Reason, α and β types, 7,32,37.
Retrospective History, 34.
Rorty Richard, 10.
Russell, Bertrand, 8.
Ryle, 5.
Sado-masochism, 36.
Science, 7,9,11,19,28,37.
Seleucus, 38.
Sensibilities, 7,9-10,18-19,21,32.
    and the Minome, 11,29,36.
    and the Gene Mover, 36.
    and Aristotle, 40.
Sensory-motor axis 15,17,18.
    and motivation, 20,29.
    and Gene Consciousness, 32.
Sexual orientation, 33.
Society, 33,37,41.
    and in animals, 22,23,26.
Socrates, 38.
Soul, 12, 39,40,41.
Stoicism, 41.

Sub-atomic dimension and energy, 34.
Teleology, 38.
Theaetetus, 38.
Theory of Mind, 33.
Tinbergen N and bird behaviour, 25.
Trajectory and perfection, 16,23,32,35.
Understanding in animals, 34,35.
Understanding, multifactorial, 4-6.
    components of, 7.
Understanding, limits of 34,36.
Understanding, Russell's definition, 8.
Understanding and the Minome, 11,13.
Unknowable, 4,6,7,8,10.
Unknowable and that which is most valuable
    in humans, 7,10,11,12,29.
    in animals, 35,36-37.
    and Aristotle, 40.
Waddington C H and canalisation, 14.
Winsatt W C, biology and culture, 30.
Wisdom, 39,40.
Woese C R, 9.
Zeno of Elea, 38.